THE
FASCINATING
ENGINEERING
BOOK

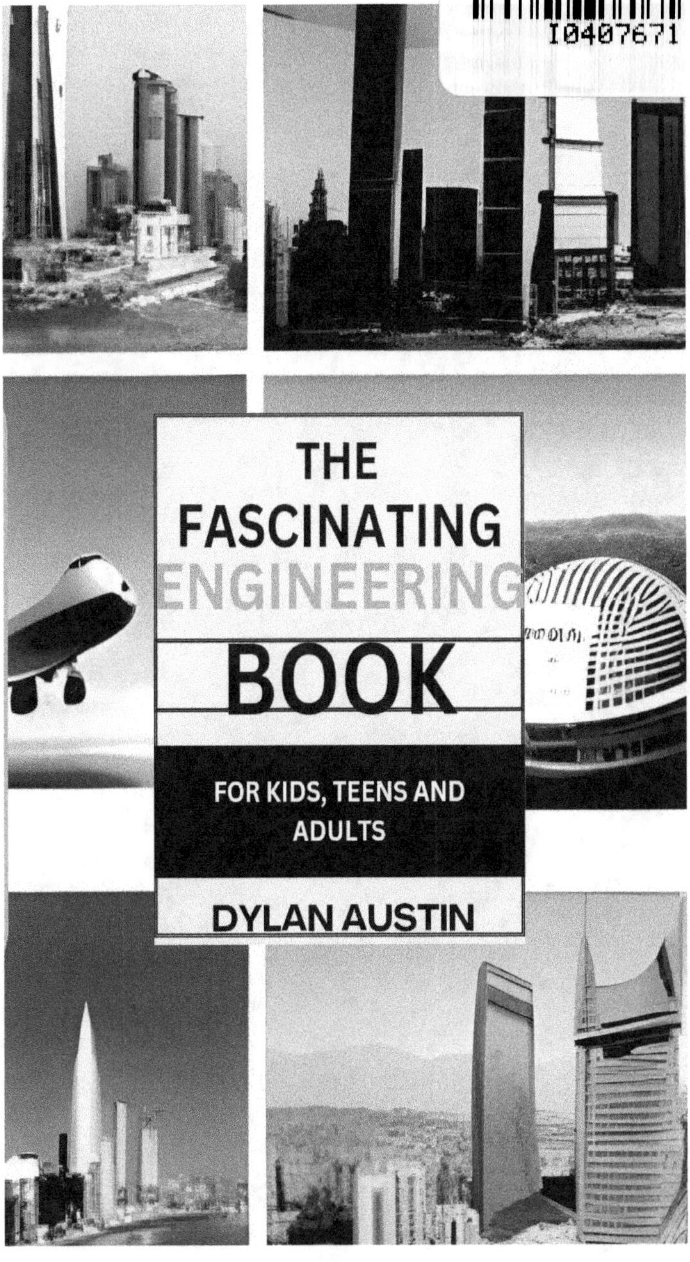

FOR KIDS, TEENS AND ADULTS

DYLAN AUSTIN

THE FASCINATING ENGINEERING BOOK

Fascinating facts for kids, teens and adults

DYLAN AUSTIN

Table of Contents

INTRODUCTION

Welcome to the beautiful world of engineering, where creativity and innovation come together to make our lives better! In this wonderful book, we invite children, teenagers and adults on an exciting journey through engineering facts that will surprise and inspire you.

From skyscrapers to complex bridges over vast rivers, engineering is the art of making dreams come true. Throughout history, engineers have used their imaginations and knowledge to create important technologies that change lives. Get ready to share amazing flavor secrets!

Did you know that the Great Chinese Wall was one of the world's most important symbols and the best engineering of its time? Imagine building a wall with a span of 13,000 miles! Speaking of the best, let's not forget the magnificence of the international space station above us. Learn about the science and technology that makes life in space possible.

But engineering is more than great monuments and research centers. This is still the power of the machines we use every day. From the smartphone in your pocket to the renewable energy that powers your home, engineers play an important role in shaping the world today. Find out how they use energy sources like wind and solar wisely to create effective solutions to our energy needs.

Have you ever wondered why skyscrapers can withstand wind and gravity? Architects create structures that appear to defy the laws of physics and reveal the secrets of architecture. Did you know that the longest and deepest tunnel in the world is also the work of engineering? Learn about the challenges engineers face when digging in mountains and underwater.

Prepare to be amazed by the technology where science fiction meets reality. Learn how engineers created man-made devices that allow humans to enter the world of bioelectronics, where technology depends on the human body.

Engineering isn't just about solving problems, it's about pushing boundaries. Marvel at the bold ideas behind record-breaking roller coasters and supersonic cars that push the limits of speed and physics. Discover the secrets of aerospace engineering and learn how airplanes and rockets defy gravity and take us to the sky and beyond.

In this book, you'll find fascinating stories of engineers who changed the world with creativity and perseverance. From historical figures like Leonardo da Vinci to modern pioneers like Elon Musk, their stories will inspire you to create, create and solve problems.

Whether you're a kid keen on creating something new, a teenager thinking about the future of engineering, or an adult who wants to see the world's perspective, this book is the gateway to the wonderful world of engineering for you. Work really hard. So turn this page and prepare to be inspired and amazed by the amazing work of our engineers and the amazing things they have yet to achieve. Your adventure in the wonderful world of technology begins now!

CHAPTER 1:

Engineering Fundamentals

- What is Engineering?

Engineering is the art of turning imagination into reality. It's an exciting cross-generational mix of creativity and science that appeals to kids, teens and adults alike.

For kids, engineering is like real-world superheroes. Build the tallest tower out of blocks, build a bridge out of popsicle sticks, and design a soaring paper plane. Through these hands-on experiences, children learn basic concepts of problem solving and critical thinking, building a foundation for lifelong innovation. Upon entering the field of engineering, teens discover the impact of engineering on their daily lives. From smartphones you can't live without to eco-friendly cars you dream of driving, engineering drives everything. Teenagers fall in love with programming, robotics and more, uncovering the secrets of technology and design. This journey introduces them to a world where their ideas can change the future.

For adults, engineering plays a role in bringing about global change. Engineers create miracles that revolutionize society, from building skyscrapers that reach into the sky to developing life-saving medical devices. A never-ending search for better, faster, and safer solutions. Adults are involved in solving complex problems, forever shaping the world by pushing the boundaries of industrial development, from energy to healthcare.

Surprisingly, engineering isn't just computation. It is also an arena for innovation. The Wright brothers designed the first airplane, and Nikola Tesla used electricity. Engineers have given us the wonders of space exploration and sophisticated nanotechnology. This is a field where creativity is not age-limited, and the solutions are as diverse as the number of people involved.

Interestingly, engineering is a global network of minds. Collaboration on different continents has resulted in the tallest bridges, fastest trains and impressive architecture. It transcends borders and cultures and unites people with a common goal: to make the impossible possible. In short, engineering engages young people, sparks their curiosity and promotes ambitious goals. It is an adventure of innovation, a journey of creation and a testament to human potential. Whether you're building sandcastles, robots, or sustainable cities, engineering is the magic that connects imagination and reality, regardless of age.

Different Types of Engineering

In the body of this chapter, we will talk about facts on the basic types of Engineering and their facts. While subsequent chapters will go into untouched types. Please note that these basic engineering types works hand-in-hand and therefore some of these facts covered in this book may likely look like they are they same. What that means is that, Civil

Engineering projects can involve those in the mechanical, Electrical, and Architectural departments.

1. Mechanical Engineering

The basic role of mechanical engineering is to design and build machines and structures that move, operate and make our lives easier. Mechanical engineers build everything from cars to airplanes, from robots to gadgets, to make the world more useful and enjoyable.

Facts about mechanical engineering

1. 3rd century BC– Archimedes Screw: Ancient Greek mathematician Archimedes invented the Archimedean Screw, a mechanical pump that carries water. It is one of the first known machines still in use today.

2. 1769 - Invention of the vacuum cleaner: James Watt developed the vacuum cleaner and changed society and business, leading the Industrial Revolution.

3. 1969 - Lunar module "Apollo 11". NASA's Apollo 11 mission, which included astronauts Neil Armstrong and Buzz Aldrin, relied on advanced technology to land the first man on the moon.

4. 1983 - First 3D printed product: Charles Hull invented stereolithography (the basis of 3D printing), which revolutionized manufacturing and prototyping.

5. 1997 - Mars Pathfinder Airbag: The Pathfinder mission uses airbags to safely land the Sojourner rover on Mars and demonstrates a new landing technique.

6. Self-correcting information: Engineers are producing materials that can self-heal, extend the life of the structure, and reduce maintenance costs.

7. Bionic limbs: Mechanical Engineering has played an important role in the development of

modern equipment that restores physical strength to the weak.

8. Hyperloop transportation: Engineers are working on Elon Musk's idea of high-speed transportation using maglev and low-speed tubes.

9. The future is connected to the elevator. A theoretical space elevator design could revolutionize space travel, using ropes to transport cargo from Earth to space stations.

10. The future belongs to medical nanorobots. Scientists are working on nanorobots, small devices for medical applications such as drug delivery and surgery in the body.

2. Civil Engineering

Civil engineering is a multidisciplinary discipline that plays an important role in the construction of today's world. Civil engineering is essentially the design, construction and maintenance of infrastructure such as bridges, roads, buildings and sewage treatment plants. Its main responsibility is to ensure the safety, efficiency and security of these structures and systems.

An interesting engineering feat is the Channel Tunnel, which connects Great Britain and France via the English Channel. This beautiful architecture involves breaking rocks under water so trains and cars can move safely.

Civil engineers also contribute to the protection of the environment. They build sustainable plumbing, use green building practices, and generate renewable energy. For example, the Delta Project in the Netherlands is one of the best projects to protect the country from flooding through dams, weirs, locks and surge barriers.

Below are some quite intriguing facts about Civil engineering.

Civil Engineering Facts

1.The Burj Khalifa structure. Burj Khalifa in Dubai, United Arab Emirates, is the tallest man-made structure in the world at 828 meters (2,717 feet).

2. Akashi Kaikyo Bridge. With a span of 1,991 meters (6,532 ft), the Akashi Kaikyo Bridge in Japan is the longest of all suspension bridges.

3. Three Gorges Dam: China's Three Gorges Dam is the largest hydroelectric dam in the world and generates large amounts of electricity while controlling floods.

4. Connecting England and France, the Canal is the longest underwater channel in the world at 50.45 miles (31.35 km).

5. Connect Hong Kong, Zhuhai and Macau. China's bridge-tunnel system is the world's longest sea bridge at 55 kilometers (34 miles).

6. High speed train network. Countries such as Japan and France have very high speed railways and trains travel at speeds in excess

of 300 km/h (186 mph).

7. Millau Viaduct. The suspension bridge in France is the highest in the world. One of the towers is 343 meters (1,125 ft) high.

8. Palm Island: A man-made island in the shape of a palm, the Palm Island in Dubai is one of the most ambitious reclamation projects in history.

9. Expansion of the Panama Canal. The expansion of the Panama Canal included the construction of new locks to accommodate larger ships and encourage international trade.

10. A Change by Bjarke Ingels: This astonishing modern architecture from Norway is a museum with a unique curvilinear structure that defies the architectural process. These achievements in civil engineering demonstrate the potential of modern architecture and design.

3. Electrical Engineering

Electrical engineering plays a key role in defining the way we live, work and communicate in the modern world. This multifaceted field is important for a number of reasons and is supported by compelling evidence highlighting its importance.

Above all, electrical engineers are architects of an interconnected world. They design, develop and maintain the infrastructure that powers our homes, businesses and industries. Power grids spanning continents and power generation technologies ranging from renewables to nuclear power provide reliable and reliable power supply. In fact, according to the US Bureau of Labor Statistics, more than 330,000 jobs were created for electrical engineers in 2020, and demand for their expertise continues to grow.

Besides, electrical engineering promotes innovation. From smartphones to spacecraft, electrical engineers are behind the cutting edge of technology. They design circuits, develop software, and optimize systems that make devices run faster, more efficiently, and smarter. Without their contributions, the technological revolution of the 21st century would not have been possible. Electrical engineers also play an important role in solving global problems. They are pioneers in the field of sustainable energy, making solar panels more efficient and wind turbines more reliable. In addition, they play an important role in developing electric vehicles, reducing carbon emissions and combating climate change.

In conclusion, we point out that electrical engineering is an indispensable element in modern society. It energizes our lives, fuels innovation and solves the most pressing problems of our time. It is a field with a bright future based on the belief that technology can change the world for the better.

Facts Surrounding Electrical Engineering

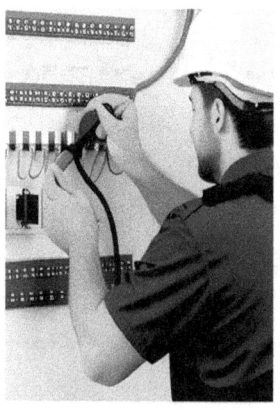

1. Wireless power transmission.
Engineers have developed a technology that allows devices to be charged wirelessly without the need for

physical connectors or cables.

2. Internet of Things (IoT). IoT devices are transforming our lives by enabling everyday objects to connect to the internet, share data, and perform tasks autonomously.

three. quantum computing. With the ability to perform complex computations at unprecedented speeds, quantum computers could revolutionize fields such as cryptography and drug discovery.

4. Carbon nanotubes. This incredibly strong and conductive material is being researched for use in next-generation electronics, from ultra-light aircraft to flexible displays.

5. Superconductors. Superconductors can conduct current without resistance, and recent advances are bringing them closer to practical applications such as highly efficient power transmission.

6. Brain-Computer Interfaces (BCIs): BCIs provide a direct connection between the human brain and external devices, opening opportunities for people with disabilities and beyond.

7. Self-healing electronics. Engineers are developing materials and circuits that can self-repair when damaged, extending the life of electronic devices.

8. Electric Vehicles (EVs): The widespread adoption of electric vehicles is transforming the automotive industry and reducing its dependence on fossil fuels.

9. Electricity from vibration. Energy harvesting technology can convert environmental vibrations, such as the sound of footsteps, into electricity, potentially powering small devices.

10. A new definition of the kilogram. In 2019, the International System of Units (SI) improved the accuracy of electrical measurements by changing the definition of the kilogram based on fundamental constants of nature, including Planck's constant. These advances reveal the ever-evolving and exciting world of electrical engineering.

4. Chemical Engineering

Chemical engineering is a multidisciplinary discipline that plays an important role in many sectors from medicine to energy production. At its core, chemical engineering is the application of chemistry, physics, biology and mathematics to design, develop and optimize processes for turning raw materials into useful products. One of the main tasks of chemical engineering is optimization. Chemical engineers analyze and improve production processes to increase efficiency, reduce costs and reduce environmental impact. They revolutionized the oil industry, for example, by developing technologies such as catalytic cracking that converts crude oil into gasoline and other valuable products. Chemical engineers are also very important in the pharmaceutical industry; They form to ensure the safety and effectiveness of chemical synthesis and purification processes. Their expertise in environmental engineering helps reduce pollution and develop clean technologies. In addition, chemical engineers are pioneers in the field of renewable energy and work to develop solar panels, fuel cells and biofuels. They make a great contribution to solving global problems such as climate change and resource depletion. In summary, the mission of chemical engineering is to transform scientific knowledge into solutions that foster innovation, sustainable development and economic growth.

Modern-day facts about Chemical Engineering

1.   For Carbon capture technology. One of the wonders of modern chemical engineering is the development of advanced carbon capture and storage (CCS) technologies that help reduce greenhouse gas emissions from industrial processes and power plants.

2. Looking at Nanotechnology in drug delivery. Chemical engineers have revolutionized medicine with nanotechnology that enables

precise drug delivery at the cellular level, resulting in more effective treatments with fewer side effects.

three. 3D printed organs. Chemical engineers are using bio-ink and 3D printing to create functional human organs that could revolutionize transplant medicine.

4. Smart materials. The research team is developing smart materials that can change physical properties in response to external stimuli, such as shape memory polymers and self-healing materials.

5. Resistant plastics. Chemical engineers are leading the way in developing environmentally friendly plastics made from renewable resources, thereby reducing dependence on fossil fuels in plastic production. 6. A breakthrough in desalination. Advances in membrane technology and chemical processes have addressed the problem of water scarcity by making desalination more efficient and cost-effective.

7. Biological purification. Chemical engineers use microbes to purify polluted environments and provide environmentally friendly solutions

to pollution problems.

8. Synthetic Biology: This emerging field combines biology and chemical engineering to design and develop biological systems for applications such as biofuel production and biopharmaceuticals.

9. Quantum computing for molecular modeling. Chemical engineers are harnessing the power of quantum computing to model complex molecular interactions and accelerate drug discovery and materials design.

10. Biological production of food. Chemical engineers are exploring the potential for biorefining of meat and other foods with the potential to revolutionize the food industry by reducing environmental impact and resource use.

5. Aerospace Engineering

Aerospace engineering is an interdisciplinary field that includes the design, construction, testing and manufacture of airplanes, aircraft and other systems. Combining elements of aerodynamics, materials science, structural analysis and antigravity kinematics, the International Space Station orbits the Earth at 28,000 miles per hour, allowing astronauts to find 16 objects, including investigating the mysteries of Earth. It captures the sunrise and sunset of the Earth's sun while surveying on the ground in microgravity.

Facts About Aerospace Engineering

1. The world's fastest manufacturer Aerospace engineering gave birth to the Parker Solar Probe, which reached speeds of up to 700,000 kilometers per hour (430,000 mph) during its mission to study the Sun, making it the fastest manufacturer. .

2. Paper Planes in Space   NASA engineers once used folded paper airplanes as part of a space shuttle microgravity experiment. These charts provide important information about the liquid on the surface.

3. Birds with different feathers The   Boeing 787 Dreamliner has wings made of carbon fiber reinforced composites that reduce weight and increase fuel efficiency, making it the first commercial airplane without metal wings.

4. Pizza in Space   Astronauts on the International Space Station (ISS) eat pizza specially prepared for eating in space. Vacuum-packed pizza is designed to prevent crumbs from floating in microgravity.

5. Parachutes Land Supersonic   on Mars Parachutes were deployed at supersonic speeds (over 1,600 mph) to slow the Curiosity rover's descent to Mars. Engineering work is required for a safe landing.

6. Airplane Decommissioning   Aerospace engineers recycle old airplanes. Decommissioned aircraft are stored and recycled in the "graveyard" at Davis-Monthan Air Force Base in Arizona; thus contributing to

sustainable development.

7. The Sound of Sonic Boom NASA's X-59 QueSST is designed to produce a "soft" sonic boom, which is a softer sound than a loud boom that can change the speed of a spacecraft.

8. Orbital Space Trash There are more than 3,000 unused satellites and a lot of space junk on Earth, and aerospace engineers are working hard to find solutions to get rid of them and reduce waste space.

9. The Sun's Interstellar Journey Aerospace engineers are investigating solar sails, light vehicles driven by solar pressure, as a possible means of interstellar travel that will enable a new way of exploring space.

10. Airplanes powered by wind turbines Scientists are exploring the possibility of using high-speed wind turbines, a unique concept in aviation power generation, to use the high pressure of the wind to generate electricity and even provide additional support to the aircraft. lb. The mystery of the sound barrier: When the plane reaches the speed of sound, noise builds up around it, causing a "sonic boom".

Overcoming this hurdle was once thought impossible, but Chuck Yeager did so in 1947, ushering in a new era in aerospace engineering.

6. Computer Engineering

Computer engineering is a part of the field that combines elements of electrical engineering and computer science to design, develop and optimize hardware and software. A dynamic place with innovators striving to make computers faster and more efficient. Did you know that the first computer viruses were actually bugs? In 1947, engineers found a moth in the relay of the Mark II Aiken relay calculator that caused the world's first computer error. This interesting fact shows his passion for expansion in computer engineering.

Cool facts about computer engineering

1. History of the first computer: The roots of computer engineering can be traced back to

the 1930s and 1940s, when the first digital electronics were produced. One of the first computers, ENIAC, weighed 30 tons and used more than 17,000 vacuum tubes.

2. Special Transistors: Transistors, the building blocks of modern computers, are getting smaller in size. In 1971 Intel released the 4004 microprocessor with 2,300 transistors. CPUs today can have millions of transistors per chip.

3. PRIVATE INTERNET: It is hidden under about 99% of the internet. This deep web often contains de-identified, proprietary and confidential information that is used for legitimate purposes such as research and government information, but there is also a

dark side to crime.

4. Quantum Computing: Quantum computers are still experimenting and using weird models of quantum mechanics. The speed with which they can process data can solve complex problems such as calculating large numbers in seconds that would take thousands of years for old computers.

5. WEIGHT OF THE INTERNET: If the weight of the Internet could be calculated, it would be heavier than a strawberry. This is due to the quality of data transmitted over the internet, stored on servers, and transmitted over fiber optic cables.

6. Silicon Valley's Extraordinary Beginning: Silicon Valley is named after the element silicon commonly used in computer chips. But in the 1970s, the name of Electronic News columnist Don Hoefler became popular, and as time passed, Silicon Valley became associated with technological development.

7. Computer Fault: The term "error" used to describe a computer error dates back to 1947, when moths caused a short circuit in the Harvard Mark II computer. Grace Hopper, a

computer scientist and military officer, is famous for calling it "debugging" after extracting errors from machines and putting them in his diary.

8. Human Brain vs. Computer: When I updated Fugaku, the world's most powerful supercomputer, in 2021, the working power of the human brain is estimated to be equal to about 2.2 exaFLOPS. It reached 442 petaFLOPs. This demonstrates the great potential of future computer engineering.

9. Internet Energy Consumption: Internet consumes a lot of energy. It is estimated that by 2020, the internet will consume as much energy as every home in Japan.

10. DNA Data Storage: Researchers have explored using DNA molecules to store digital data due to their incredible data density. It's possible to store the entire internet in just a few grams of DNA!

7. Industrial Engineering

Industrial engineering is a dynamic field that combines creativity, problem solving and technical skills to develop complex systems. An interesting fact about the engineering industry is its role in the industrial revolution. For example, in the early 20th century, industrial engineers like Frederick Taylor pioneered scientific management techniques that increased productivity and changed factories.

Another interesting aspect of Industrial Engineering is its wide application area. These professionals are not in a single industry; They use their skills in every field, from manufacturing to healthcare, from logistics to finance. This shift enables business professionals to play an important role in building the world's infrastructure and systems.

Industrial engineers are also at the forefront of innovation. They use new technologies such as data analytics and artificial intelligence to improve processes and improve decision making. This technology is not only efficient, but also supports sustainable practices that reduce waste and energy consumption. Industrial engineering is essentially the driving force behind the optimization of almost every system we encounter, making it an important discipline in today's world.

Whether building a more efficient supply chain or improving health, industrial engineers are the architects of better production and a better future.

Facts surrounding Industrial Engineering

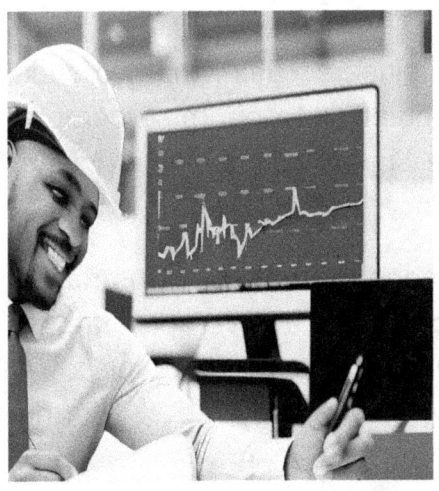

1.   Frank Gilbreth, one of the pioneers of industrial engineering, is credited with creating the first simple project called

Motion Study" It also inspired the book and movie "Cheaper by the Dozen."

2. First introduced by Toyota in the 1950s, the concept of "just-in-time" (JIT) manufacturing that minimizes inventory and waste is a cornerstone of modern industrial engineering.

3. Industrial engineers often use a process called "human factor engineering" to design products and systems suitable for human use, including thinking, body and ergonomics.

4. The "Pareto" rule, often referred to as the 80/20 rule, is often used in design. This means that around 80% of the results are 20% of the cause, which helps identify key areas for improvement.

5. Industrial engineers use mathematical modeling and simulation to optimize complex systems such as supply chains for efficiency and cost reduction.

6. During the Second World War, industrialists developed Project Lilliput, a covert program for small military equipment for use in covert operations, which spurred further

developments in microtechnology.

7. Time and motion studies by industrial engineers can be made more efficient by identifying and eliminating unnecessary and unprofitable activities. Day

8. The concept of Six Sigma is derived from design principles to achieve the best through reduction and change in the process.

9. In the 20th century, industrial engineers worked on movement to improve work efficiency, which led to the development of modern assembly tools.

10. Engineers have helped companies like Amazon deliver quality packages by developing sophisticated algorithms to optimize supply chain logistics.

8. Environmental Engineering

Environmental engineering is a branch in engineering that uses scientific methods and engineering skills to come up with new solutions to protect, conserve and restore the environment. Interestingly, environmental engineers helped build the Great Green Wall of Africa, a major tree planting project to combat desertification and other issues, proving these issues play an important role in the world.

Interesting facts surrounding environmenntal

1. Green roofs: They often design and use green roofs that cover plants in cities to reduce energy consumption, improve air quality and reduce heat impact.

2. Wastewater treatment: They play an

important role in the development of technological technologies for treating wastewater and recycling water and reducing the freshwater level.

3. Renewable energy: Environmental engineers are responsible for the development of renewable energy such as wind farms, solar farms and dams to promote sustainable energy production.

4. Air Quality Control: They work on the design and control of air pollution control systems, as well as the creation of washing machines and water filters to reduce air pollution.

5. Land remediation: Environmental engineers are concerned with cleaning up contaminated areas using techniques such as biological remediation and soil vapor extraction to return the land to a safe environment.

6. Climate Change Mitigation: They contribute to climate change by researching carbon capture and storage (CCS) technologies to reduce carbon monoxide emissions from industrial processes.

7. Sustainable buildings: Environmental

engineers help create environmentally friendly buildings with energy-saving systems, sustainable materials and good waste management.

8. Environmental Impact Assessments: They conduct comprehensive assessments to assess the potential impact of a project on the environment, ensure regulatory compliance, and minimize ecological risks.

9. Water resource management: Environmental engineers develop strategies for sustainable water management, including the construction of water reservoirs, water supply systems, and flood protection.

10. Ecosystem Restoration: They strive to restore damaged ecosystems such as rivers and forests to their natural state by promoting biodiversity and ecological balance.

9. Biomedical Engineering

This exciting discipline uses elements from various branches of engineering, including electrical, mechanical and chemical, to develop innovative solutions to complex problems in medicine and biological research. Biomedical engineering is essentially about improving human health and well-being through the development of technologies, medical devices and treatment strategies. The potential of this field is enormous and its applications cover almost all aspects of modern medicine and therapy.

Define Biomedical Engineering:

Biomedical Engineering, often abbreviated as BME, is a field dedicated to applying engineering principles and techniques to solve problems related to biology and medicine. It bridges the gap between two seemingly disparate fields and plays an important role in improving our understanding of the human body, diagnosing diseases, regenerating and healing entire systems.

Fun Fact: One of the most fascinating aspects of biomedical engineering is its ever-evolving nature. As our knowledge of biology and medicine has grown, so has our knowledge of biomedical engineering. Through continuous innovation, things once thought impossible can be achieved. For example, the idea of 3D printing for organ or tissue transplantation, once considered science fiction, is now becoming a reality at BME.

Biomedical Engineering Achievements:

1. Medical Imaging Advances: Biomedical engineers have played an important role in the development of advanced medical equipment such as MRI, CT scans and ultrasound, revolutionizing Medical Imaging Technology. diagnosing and treating disease.

2. Prosthesis: This field has contributed to the development of advanced prostheses with complex control systems, enabling the disabled to regain strength and flexibility.

3. Biomechanics Research: Biomedical engineers study the mechanics of the human body, leading to innovations in the design of exercise equipment, injury prevention, and rehabilitation.

4. Drug Delivery Systems: They create targeted drug delivery systems that allow the administration of drugs and reduce side effects.

5. Electronic Products: BME makes progress in the development of artificial organs such as liver and kidney, offering life-saving solutions to patients waiting for change.

6. Genome Sequencing: BME-driven advances in genomics are advancing the path to personalized medicine by tailoring treatments to an individual's genetic makeup.

7. Neuroprosthesis: Biomedical engineers have developed brain-computer interfaces that allow paralyzed people to control computers and robotic devices with their thoughts.

8. Tissue Engineering: This field holds promise for organ transplantation and disease modeling by enabling the growth of tissues and organs in the laboratory.

9. Rehabilitation Equipment: BME develops new rehabilitation equipment such as exoskeletons and assistive devices to improve the quality of life of people with disabilities.

10. Biosensors: Biomedical engineers are revolutionizing healthcare by developing precision biosensors for early detection of disease, monitoring and surveillance of health. Continuous evaluation is very important.

In summary, biomedical engineering is a dynamic and rapidly evolving field that seamlessly blends engineering models with complex biology and medicine. Their achievements have revolutionized healthcare, from technology to life-saving medical devices, personalized medicine and more. As technology continues to advance, the potential of biomedical engineering to improve health and improve the human body is limitless, making it one of the most promising disciplines and useful in today's world.

10. Software Engineering

Software engineering is the discipline that forms the backbone of our modern connected world. Among its students is a systematic and scientific approach to the design, development, testing and maintenance of software systems. But this seemingly simple interpretation hides great influence and interesting facts about the field.

Imagine you are in a world without software engineering. Your smartphone would just be a piece of plastic and metal that cannot perform many functions. The Internet would have been a distant dream, and the idea of driverless cars or augmented reality would have have been pushed into the realm of science fiction. But software engineering has emerged as a magic that transforms hardwares into devices that can be connected, calculate and design.

Now let's examine some Interesting Facts About Software Engineering:

1. Birth of a discipline: Software engineering as a formal discipline is a term coined during the 1968 NATO Software Engineering Conference. Before that, programming was bad business.

2. Margaret Hamilton Code: The software for NASA's Apollo 11 mission (which sent the first man to the moon) was created by Margaret Hamilton. Its code is so powerful that it prevents the process from being overloaded during the moon landing.

3. Y2K error: As the year 2000 approached, there was widespread concern that computer systems would malfunction due to the date virus. Software engineers around the world are trying to avoid the "Y2K" crisis and ensure a smooth transition into the new millennium.

4. Open source movement: The open source

software movement, represented by projects such as Linux, Apache, Mozilla, revolutionized the software industry. This is a testament to software engineering collaboration.

5. Moore's Law: Although not directly related to software engineering, Moore's Law predicts that the number of transistors in a microchip will double every two years, triggering rapid growth in software capabilities.

6. World Wide Web: Created by Tim Berners-Lee in 1989, the World Wide Web has revolutionized the way we access and share information, forming the basis of countless software applications. .

7. Agile Methodologies: Agile software development methodologies that emphasize flexibility and customer collaboration have become the industry standard.

8. Mobile Innovation: The rise of smartphones and mobile apps has led to an explosion in software development over the past year. This change has created a brand new business and business model.

9. Artificial intelligence and machine learning:

Software engineering is at the forefront of the development of artificial intelligence and machine learning algorithms by enabling applications such as virtual assistants, strategies and self-driving cars.

10. Quantum computing: Quantum computing, still in its infancy, promises to revolutionize software engineering by solving complex problems at speeds unimaginable with classical computers.

Software engineering is actually the magic wand that turns the virtual into reality. It combines creativity and efficiency to turn ideas into lines of code and digital knowledge. It is a discipline that is constantly evolving and adapting to new challenges and opportunities. Standing on the abyss of an increasingly digital future, software engineering remains at the heart of innovation and is poised to shape the world in ways we can't yet imagine.

CHAPTER 2:

Nature-Inspired Engineering

- Biomimicry: Learning from Nature

Often referred to as "nature inspired design", biomimicry is a new field of engineering inspired by the complex patterns, processes and mechanisms found in nature. By studying and modeling biological solutions, engineers and scientists are working to create new technologies that are useful in a variety of fields. This approach not only considers the environment, but also offers solutions to complex problems. This article provides an in-depth look at the world of engineering biomimicry, highlighting its principles, applications, and future developments.

Principles of Engineering Bionics

The key to biomimicry is the study of the nature of experimental design and techniques. Which includes three principles.

1. Emulation of nature. Architects study patterns and structures such the aerodynamics of bird wings, honeycomb structures to create good and functional designs. For example, the design of today's wind turbine was inspired by the shape of humpback whale fins to provide better energy.

2. Simulation: Nature is the master of efficiency. Engineers create new systems by observing and repeating biological processes. A famous example is Velcro, which arises from thorns in feathers.

3. Ecosystem Access: Natural ecosystems are models for resilience. Engineers seek to create systems that live in harmony with their environment, similar to natural ecosystems.

Applications of Bionics in Engineering

Bionics has many applications in various fields of engineering.

4. Aviation and Space Industry: The study of bird flight has allowed more aircraft to be built, reducing fuel consumption and emissions.

5. Materials Science: Scientists create materials inspired by powerful spiders having effective application in medicine and architecture.

6. robots. Bionic robots can traverse rough terrain or perform small tasks by mimicking the movement and behavior of animals.

7. Energy: Leaf-style solar panels maximize energy capture, and palm-shaped wind turbines maximize wind power.

8. architectural. Buildings designed with biomimetic structures can reduce energy consumption by automatically adjusting the temperature and lighting.

Future Destinations

As technology advances, the field of engineering biomimicry will evolve as well. Future developments may include:

1. Biocomputer: Designing computers and algorithms that mimic the operation of the brain's neural networks.

2. Treatment: Development of bio-based medicine to cure disease.

3. Sustainability: Extending biomimicry principles to urban planning and infrastructure to create strong and resilient cities.

4. Space Exploration: Build spaceships and habitats that recreate Earth's ecosystems to support long-term operations.

- Engineers Inspired by Animals and Plants

Below are 6 inventions inspired by Animals and plants. These inventions includes:

1. Velcro Inspiration: Swiss engineer George de

Mestral was inspired by a thorn in his dog's fur while hiking. This led to the invention of Velcro, a fastening system based on tiny hooks and loops that revolutionized industries from clothing to space exploration.

2. Shinkansen beak design. The design of the Japanese Shinkansen was inspired by the kingfisher's beak. Engineers mimicked its streamlined shape to reduce noise and increase speed, making it one of the fastest trains in the world.

3. The impact of the aerodynamics of the Eiffel Tower. Gustave Eiffel, the engineer of the Eiffel Tower, was inspired by the aerodynamic shapes of the natural world. The tower's lattice structure is remarkably similar to that of a bird's bone structure, making it light and strong.

4. Sharkskin-like surface: Engineers have developed materials and finishes inspired by the unique texture of sharkskin. These biomimetic surfaces reduce drag and inhibit bacterial growth, which has applications in swimwear, ship hulls and medical devices.

5. The Lotus effect of self-cleaning surface. The lotus flower's ability to repel water and

stay clean in dirty environments inspired engineers to create self-cleaning surfaces. This technology is currently used in building and automotive paint to maintain cleanliness and reduce maintenance costs.

6. Needleless injection. Mosquitoes have evolved needle-like proboscis to pierce the skin and feed on blood. Inspired by this concept, engineers developed a needleless injection system. The device uses high-pressure air or other methods to painlessly deliver medication through the skin, reducing the need for conventional needles.

CHAPTER 3:

Incredible Structures

- Skyscrapers: Reaching for the Sky

1. The Lotte World Tower, Seoul, South Korea.
Construction date: Started in 2011, completed in 2017.
Situation: Lotte World Tower was built as a Korean landmark and an integral part of the larger Lotte World complex. The 555m tall structure is a testament to Korea's economic growth and architectural innovation.

2. Shanghai Tower, Shanghai, China (2015):

Opened in 2015, Shanghai Tower is 632m high. It was built to address the city's growing population density and promote sustainable urban living. The tower's distinctive, winding design was designed by architecture firm Gensler.

3. One World Trade Center, New York, USA (2014): The World Trade Center was built in 2014 to replace the original World Trade Center which was destroyed on September 11, 2001. It is 541m tall and is a symbol of resilience and regeneration. SOM architect David Childs designed the tower.

4. Taipei 101, Taipei, Taiwan (2004): Taipei 101 held the title of tallest building in the world from 2004 to 2010. The tower was built in response to Taiwan's frequent seismic activity and typhoons. Designed by C.Y. Lee & Partners reaches a height of 508 m.

5. Petronas Towers, Kuala Lumpur, Malaysia (1998): Completed in 1998, the Petronas Towers were once the tallest building in the world (452 metres). Their construction symbolized Malaysia's economic growth and ambition. Designed by César Pelli, the tower has a distinctive Islamic design.

1. Akashi Kaikyo Bridge (Japan):

- Date: Completion in 1998.

- Situation: Akashi Kaikyo Bridge, also known as Jinju Bridge, was built to connect Japan's Awaji Island and Honshu Island. Because it is located in a seismically active area and is prone to typhoons, it faced numerous engineering challenges. The completion of the bridge greatly improved transport links between Japan's two main islands.

2. Golden Gate Bridge (USA):

- Year: Completion in 1937.

- Situation: The Golden Gate Bridge in San Francisco, California was built as a public project during the Great Depression. The ship has become an iconic symbol of the city and a testament to its engineering ingenuity as it traverses the complex Golden Gate Strait with strong currents and frequent fog.

3. Sydney (Australia) Harbor Bridge:

- Year: Completion in 1932.

- Situation: The Sydney Harbor Bridge is an important transport link linking Sydney's CBD with the northern suburbs. Its construction provided jobs during the Great Depression and was seen as a symbol of hope. The opening during the Great Depression helped boost the city's economy.

4. Viaduct Millau (France):

- Date: Completion in 2004

- Situation: One of the tallest cable-stayed bridges in the world, the Millau Viaduct was built to reduce traffic congestion in the region and improve transport links between Paris and the Mediterranean coast. Innovative design and construction, including the use of temporary towers, minimized environmental impact during construction.

5. Danyang-Kunshan Bridge (China):

- Date: Completion in 2010

- Situation: The Danyang-Kunshan Bridge is part of the Beijing-Shanghai high-speed rail and is one of the longest viaducts in the world. It was built to expand China's high-speed rail network and facilitate travel between Beijing and Shanghai. Its completion was an important milestone in China's infrastructure development.

- Underwater Wonders: Submerged Engineering

1. Japan Atlantis - Yonaguni Monument

- Date: Discovered in 1986.

- Situation: The Yonaguni Monument, located off the coast of Yonaguni Island in Japan, was discovered by a diver in 1986. Some believe that these huge underwater rock formations are man-made structures similar to ancient ruins. However, its origins are still the subject of controversy among scholars.

2. Lost Egyptian City of Tonis-Heraklion

- Date: Found in 2000

- SITUATION: In 2000, the ancient Egyptian city of Tonis Heraklion was discovered submerged in the Mediterranean Sea. It disappeared over a thousand years ago due to natural disasters and rising sea levels. The ruins of the city provide valuable information about the culture and trade of ancient Egypt.

3. The Lion City of the Submerged Qiandao Lake

- Date: 1959 self-immolation.

-Circumstances: In 1959, the Chinese government intentionally flooded the ancient city of City Hall, also known as the Lion City, to create Qiandao Lake. Built during the Ming and Qing dynasties, the city is remarkably well-preserved underwater and has become a popular diving destination.

4. The sunken battleship USS Arizona

- Date: Sunk on December 7, 1941

- SITUATION: US Navy battleship USS Arizona was sunk during the attack on Pearl Harbor on

December 7, 1941. She is partially submerged in port as a memorial to the 1,177 sailors who died. their life. It is a vivid reminder of the events that led America into World War II.

5. MUSA Underwater Sculpture

- Date: Installation started in 2009.

- Situation: The Mexican Underwater Museum (MUSA) started creating an underwater sculpture garden in 2009. Located off the coast of Cancun and on Isla Mujeres, these sculptures serve as artificial reefs and promote marine life while providing divers and snorkelers with a unique and surreal experience.

CHAPTER 4:

Exploring Space and Beyond

- The Journey to Outer Space

1. Mysterious Fast Radio Burst (FRB).  Astronomers have detected powerful and unexplained radio bursts in deep space known as Fast Radio Bursts. These intense, millisecond-long signals originate from distant galaxies, and their

origins remain a fascinating cosmic mystery.

2. The dominance of dark matter. Recent observations have shown that dark matter, a mysterious substance that does not emit, absorb, or reflect light, makes up about 85% of the matter in the universe. The discovery challenges our understanding of the universe, where we are still trying to directly detect or fully understand dark matter.

3. Solar ghost neutrino. Solar neutrinos, elusive subatomic particles created by nuclear reactions in the sun, have been found to oscillate between different flavors during their journey to Earth. This discovery changed our understanding of particle physics and has influenced the entire field of astrophysics.

4. Supermassive black hole duo: Astronomers have discovered a pair of supermassive black holes in the process of merging. When these cosmic giants eventually collide, they create powerful gravitational waves that may help us better understand the nature of gravity itself.

5. Acceleration of the expanding universe. Surprisingly, the universe is not only expanding, but the rate of expansion is also accelerating.

Dark energy, a substance far more mysterious than dark matter, is believed to be responsible for this acceleration. This discovery challenges our fundamental understanding of the universe and its fate.

6. Mysterious Fast Radio Bursts (FRBs): Scientists have identified the cause of fast radio bursts in space, intense and brief radio bursts that appear to originate from magnetars, neutron stars with incredible magnetic strength. The discovery raised interesting questions about the extreme conditions under which these explosions occur.

7. A spinning black hole. It was believed that there are two types of black holes: rotating and non-rotating. However, recent observations have shown that some black holes exhibit frame dragging, a phenomenon that allows them to rotate at nearly the speed of light, causing deep distortions in space-time.

8. Mysterious interstellar objects. In 2017, an unusual interstellar object called 'Oumuamua passed through our solar system. Its odd cigar-shaped shape and choppy acceleration defy existing explanations, leading to speculation about its true nature, including possibly an

extraterrestrial probe.

9. A star that never ages. Judging by their

estimated ages, some stars appear to be older than the universe itself. This paradox challenges our understanding of stellar evolution and the age of the universe, and suggests that there may be gaps in our knowledge of how stars form and evolve.

- Space Stations and Habitats

1. Space stations such as the International Space Station (ISS) are excellent examples of international cooperation. Many countries, such as the United States, Russia, Europe, Japan,

and Canada, have contributed to its construction and operation.

2. Microgravity research.

Interestingly, the space station provides a unique environment for scientific research. Microgravity allows scientists to conduct experiments in the fields of physics, biology, and materials science that cannot be doane on Earth.

3. Long missions. Astronauts on the space station can stay in orbit for months. The longest stay on the ISS lasted nearly year, which helped researchers study the effects of long-duration spaceflight on the human body.

4. Life support systems. The space habitat is equipped with a state-of-the-art life support system that recycles air and water, allowing it to be self-sustaining for long periods of time. This technology will be essential for future long -range missions to the Moon or Mars.

5. Spacewalk. Astronauts often perform spacewalks (spacewalks) to repair and maintain space stations. These risky but important operations are essential to business

operations.

6. International Institute. The ISS is the only laboratory in orbit. It hosts experiments in various fields, including medicine, biology, astronomy, and earth science.

7. Prelude to Deep Space Exploration: Space habitats and stations are critical to preparing for deep space missions. They help develop the skills and experiences needed for future travel to the Moon, Mars and beyond.

8. Space tourism. The development of commercial space stations and habitats is already on the horizon. Companies like SpaceX and Blue Origin plan to offer people the opportunity to visit and stay in space for both entertainment and research purposes.

9. Eternal sunrise and sunset. Astronauts constantly witness sunrises and sunsets, experiencing unique phenomena. This is because they orbit the earth about every 90 minutes, so we see the sun rise in the east and set in the west several times a day, unlike the earth where there is only one sunrise and one sunset per day.

10. weightless. In space, astronauts experience microgravity, which means floating freely inside a spacecraft. This is very different from Earth where gravity keeps us on the ground. This weightlessness can have various effects on the human body, so astronauts need to exercise regularly to counter the loss of muscle and bone mass. three. temperature extremes. Space can be very hot or very cold. Under direct sunlight, the temperature can reach over -250 degrees Fahrenheit (120 degrees Celsius), but in the shadow of a cosmic ray the temperature can drop to -250 degrees Fahrenheit (-157 degrees Celsius). Astronauts must wear special clothing and spacecraft to protect themselves from rapid temperature changes.

11. There are 16 sunrises and sunsets per day. Due to its orbital speed, astronauts aboard the International Space Station (ISS) see about 16 sunrises and sunsets each day. This rapid cycle can disrupt your circadian rhythm, so it's important to stick to a strict schedule to maintain your day and night mood.

12. Lack of seasons. Unlike Earth, there are no seasons in space. On our planet, the change of seasons is caused by the tilt of the Earth's axis

rotating around the Sun. Because there is no such axis tilt in space, astronauts do not experience seasons the way we do on Earth. Instead, it is in a constant and unchanging environment, except for day and night changes due to orbital motion.

CHAPTER 5

The inspiring stories of the Ingenious Minds in the spotlight.

Leonardo Da Vinci: A Genius and His Intriguing Inventions

Leonardo da Vinci was a man of many talents. He was born in Vinci, Italy in 1452 and went on to become one of history's greatest geniuses. Leonardo's insatiable curiosity led him to explore a wide range of fields, from art to technology, from anatomy to astronomy. But what really set him apart were his ingenious inventions, many of which were centuries ahead of their time.

One of Leonardo's most interesting inventions was the flying machine. At a time when the idea of human flight was nothing more than a pipe dream, Leonardo sketched and designed several flying devices. His most famous development was the ornithopter, a flapping machine inspired by bird flight. Leonardo understood that imitation of nature is the key to flight. Although he never built an actual model of an onytopter himself, his drawings and ideas paved the way for future aviation pioneers such as the Wright brothers.

For kids, the story of Leonardo's flying machine is a story of determination and imagination. It teaches that with the right combination of creativity and perseverance, even the wildest dreams can come true. Adults, on the other hand, can understand the foresight behind Leonardo's designs. He just didn't want to fly. He wanted to understand the principles of flight that continue to drive innovation in aviation to this day.

Leonardo's inventions were not limited to the sky. He also had a profound influence on underwater exploration. In the 15th century, he developed the wetsuit, the forerunner of modern diving equipment. His invention was a leather suit with a helmet and a breathing tube that rose to the surface. It wasn't as perfect as today's scuba gear, but it was an innovative concept that allowed divers to explore the depths of the ocean.

For children, Leonardo's wetsuit is a testament to the human desire to explore the unknown even when faced with great challenges. This reminds us that curiosity knows no bounds. Adults will admire Leonardo's ability to bridge the gap between imagination and practicality and lay the groundwork for future advances in underwater research and technology.

Another interesting invention of Leonardo was the armored tank. At a time when knights on horseback were leading the war, Leonardo envisioned a vehicle capable of withstanding enemy fire and transporting soldiers into battle. Its design was a round tank-like structure with cannons mounted on all sides. No tanks were built during his lifetime, but he foreshadowed advances in military technology.

The story of Leonardo's armored tank teaches children the importance of adaptability and innovation in the face of changing circumstances. It also shows that creativity can be a driving force for progress even in times of conflict. For adults, this is a reminder of the ongoing human need to improve existing technology, a drive that has led to innovations in military strategy and armored vehicles.

The life and inventions of Leonardo da Vinci continue to inspire generations of children and adults. His insatiable curiosity, boundless imagination and determination to bring his ideas to life are timeless examples of the power of human creativity and innovation. Flying machines, diving suits and armored tanks – Leonardo's legacy remains a testament to the limitless possibilities of the human spirit.

Elon Musk: Pioneer of Innovation

A man synonymous with innovation and bold ventures, Elon Musk has captured the world's imagination with his groundbreaking inventions. From electric cars to reusable rockets, Musk's journey is a testament to human ingenuity and determination.

Musk's passion for technology and innovation began at an early age. He was born in South Africa in 1971 and showed an aptitude for programming from an early age, creating his first video game at just 12 years old. This early passion for technology laid the groundwork for his future endeavors.

One of Musk's most exciting inventions is the Tesla electric car. In 2004, he co-founded Tesla

Motors with a mission to accelerate the world's transition to sustainable energy. The Tesla Roadster, launched in 2008, was a milestone as it was the first electric sports car to reach a mainstream audience. Musk's vision of clean electric transportation challenges the status quo and pave the way for the subsequent electric vehicle revolution.

Tesla's Model S, launched in 2012, was another revolutionary moment. He showed that electric vehicles can be environmentally friendly and luxurious. The Model S' long-distance driving capability and impressive acceleration set a new standard for electric vehicles, making it more popular with more consumers.

But Musk's innovations weren't limited to roads. In 2015, he launched the Powerwall, a home battery system designed to store energy from solar panels or the grid. This invention represents an important step towards making renewable energy more affordable and reliable in homes around the world. The Powerwall reduces dependence on fossil fuels by storing surplus energy during the day and using it at night.

Musk's desire to explore and colonize Mars

also spurred his efforts. Founded in 2002, SpaceX sought to reduce the cost of space transportation and make colonization of other planets possible. One of SpaceX's most exciting developments is the reusable Falcon 9 rocket. This innovation could revolutionize space travel by significantly reducing the cost of launching payloads into space.

In 2020, SpaceX hit another milestone with the successful launch and return of astronauts to and from the International Space Station aboard the Crew Dragon spacecraft. Musk's dream of making space travel more accessible to ordinary people is becoming a reality, and exploration of Mars is planned in the not-too-distant future.

Musk's adventures go beyond Earth and Mars. In 2016, he founded Neuralink, a company specializing in the development of brain-computer interfaces. The goal of this fascinating invention is to combine the human brain with artificial intelligence, potentially unlocking a new level of human capabilities. Neuralink is still in the early stages of development, but promises to revolutionize the way we interact with technology.

Elon Musk's path as an inventor and entrepreneur is simply outstanding. His ability to foresee and realize innovative technologies has transformed the industry and pushed the boundaries of what is possible. Musk's inventions, including electric cars, reusable rockets and brain-computer interfaces, continue to captivate the world and inspire the next generation of innovators. His legacy will undoubtedly be remembered as one of the most exciting chapters in the history of technology and space exploration.

CONCLUSION

In the realm of human achievement and innovation, the pages of this fascinating engineering facts book have unfolded a world of wonder and amazement. From the towering skyscrapers that pierce the heavens to the intricate machinery that powers our daily lives, engineering has woven a tapestry of marvels that have reshaped the course of history.

As we draw the final curtain on this captivating journey through the annals of engineering, we are reminded that human ingenuity knows no bounds. The stories within these pages have unveiled the relentless pursuit of knowledge, the unyielding spirit of creativity, and the unwavering dedication of engineers who have transformed dreams into reality.

From the awe-inspiring feats of ancient civilizations to the cutting-edge technologies of the present day, we have witnessed the evolution of engineering as a driving force behind progress. He has crossed chasms, crossed seas, and surpassed what was once thought impossible.

The facts and stories contained in this book testify to the power of human collaboration and the relentless pursuit of excellence. They remind us that every bridge, every building, every gadget is a testament to the tireless efforts of those who dare to dream and build.

In the world of engineering, every challenge is an opportunity, and every problem is a puzzle waiting to be solved. As we close this page, let us take the inspiration to continue pushing the boundaries of what we can achieve, question the status quo, and embrace the limitless possibilities that lie ahead.

In a world where innovation is the currency of progress, engineers are the architects of change. So whether you're an aspiring engineer, a seasoned professional, or just a curious reader, let this book be your beacon of inspiration and remind you that the world of engineering is a realm of endless fascination where the sky is the only limit.

Finally, we hope that you will continue to explore, innovate and dream, eternally fascinated by the wonders of engineering, because the future of our world is in the hands of those who want to create a brighter future.